Name: _____

# 11+

# Non-Verbal Reasoning

# Workbook
## Age 8 – 10

**Peter Francis**

GALORE PARK

AN HACHETTE UK COMPANY

**Orders: Teachers,** please contact Hachette UK Distribution, Hely Hutchinson Centre, Milton Road, Didcot, Oxfordshire OX11 7HH. Telephone: (44) 01235 400555. Email: primary@hachette.co.uk. Lines are open from 9 a.m. to 5 p.m., Monday to Friday.

**Parents, Tutors** please call: (44) 02031 226405 (Monday to Friday, 9:30 a.m. to 4.30 p.m.). Email: parentenquiries@galorepark.co.uk

Visit our website at www.galorepark.co.uk for details of other revision guides for Common Entrance, examination papers and Galore Park publications.

ISBN: 978 1 4718 4934 3

© Hodder & Stoughton 2016

First published in 2016 by

Hodder & Stoughton Limited

An Hachette UK Company

Carmelite House

50 Victoria Embankment

London EC4Y 0DZ

Impression number    13  12  11  10  9  8

Year        2025  2024  2023

Illustrations by Peter Francis

Typeset in India

Printed in the UK

A catalogue record for this title is available from the British Library.

# Contents and progress record

Use these pages to record your progress. Colour in the boxes when you feel confident with each skill and note your scores for the 'Test yourself' and workout questions.

# How to use this workbook

## Introduction

This workbook has been written to help you develop your skills in Non-Verbal Reasoning. The questions will help you:

- learn how to answer different types of questions
- build your confidence in answering these types of questions
- develop new techniques to solve the problems easily
- practise maths skills that can help improve your abilities in Non-Verbal Reasoning
- build your speed in answering Non-Verbal Reasoning questions towards the time allowed for the 11+ tests.

## Pre-test and the 11+ entrance exams

The Galore Park 11+ series is designed for pre-tests and 11+ entrance exams for admission into independent schools. These exams are often the same as those set by local grammar schools too. 11+ Non-Verbal Reasoning tests now appear in different formats and lengths and it is likely that if you are applying for more than one school, you will encounter more than one of type of test. These include:

- pre-tests delivered on-screen
- 11+ entrance exams in different formats from GL and CEM
- 11+ entrance exams created specifically for particular independent schools.

Tests are designed to vary from year to year. This means it is very difficult to predict the questions and structure that will come up, making the tests harder to revise for.

To give you the best chance of success in these assessments, Galore Park has worked with 11+ tutors, independent school teachers, test writers and specialist authors to create this series of workbooks. These workbooks cover the main question types that typically occur in this wide range of tests.

### For parents

This workbook has been written to help both you and your child prepare for both pre-test and 11+ entrance exams.

The content doesn't assume that you will have any prior knowledge of Non-Verbal Reasoning tests. It is designed to help you support your child with simple exercises that build knowledge and confidence.

The exercises on the **learning spreads** can be worked through either with your support or independently. They have been constructed to help familiarise your child with how a question type works in order to build confidence in tackling real questions.

The **maths workout** sections are provided to help consolidate learning in related areas of maths.

## For teachers and tutors

This workbook has been written for teachers and tutors working with children preparing for both pre-test and 11+ entrance exams. The wide variety of question types is intended to prepare children for the increasingly unpredictable tests encountered, with a range of difficulty developed to prepare them for the most challenging paper and on-screen adaptable tests.

## Working through the workbook

- The **contents and progress record** helps you keep track of your progress. Complete it when you have finished one of the **learning spreads** or **maths workout** sections.
  - Colour in the 'Completed' box when you are confident you have mastered the skill.
  - Add your 'Test yourself' scores to track how you are getting on and to work out which areas you may need more practice in.
- **Chapters** link together types of questions that test groups of skills.
- **Learning spreads**, like the one shown here, each cover one question style.

**Have a go**

Try these activities to build your skills towards answering the exam-style questions.

**Test yourself**

Complete a set of exam-style questions that includes some challenging problems.

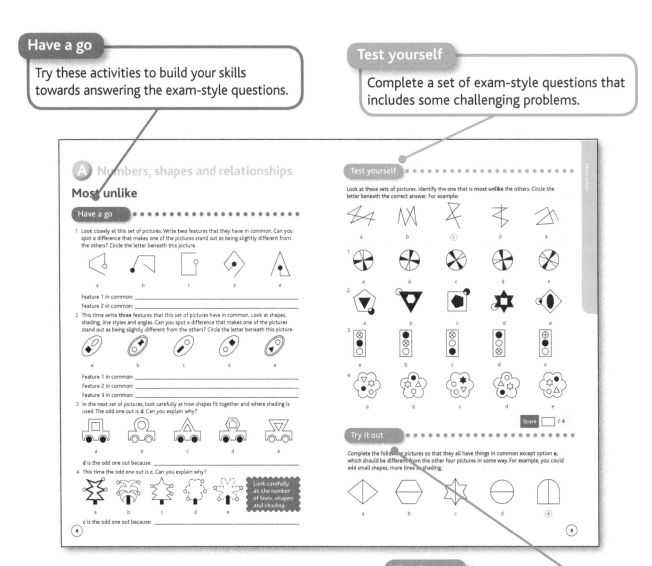

**Try it out**

Use your new skills to create your own questions or complete a fun activity.

- **Maths workouts** help you to practise familiar skills that link to the Non-Verbal Reasoning questions in this workbook.
- **Answers** to the **Have a go**, **Test yourself** and **Try it out** questions can be found in the middle of the workbook. Try not to look at the answers until you have attempted the questions yourself. Each answer has a full explanation so you can understand why you might have answered incorrectly.

## Test day tips

Take time to prepare yourself the day before you go for the test: remember to take sharpened pencils, an eraser and a watch to time yourself (if you are allowed – there is usually a clock present in the exam room in most schools). Take a bottle of water in with you, if this is allowed, as this will help to keep your brain hydrated and improve your concentration levels.

... and don't forget to have breakfast before you go!

# Continue your learning journey

When you've completed this workbook, you can carry on your learning right up until exam day with the following resources.

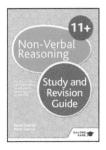 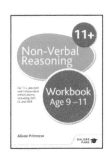 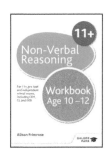  

The *Study and Revision Guide* introduces the skills and question types you may encounter in your Non-Verbal Reasoning 11+ entrance exams. Questions are broken down into the familiar maths areas you have learned about in school, with full explanations to show how these work together in the Non-Verbal Reasoning questions.

The workbooks will develop your skills, with many practice questions. To prepare you for the exam, these books include even more question variations that you might encounter. The more question types you practise, the better equipped for the exams you'll be. All the answers are explained fully.

*Workbook Age 9–11*: Experiment with further techniques to improve your accuracy.

*Workbook Age 10–12*: Develop fast response times through consistent practice.

The *Practice Papers* (books 1 and 2) contain four training tests and nine model exam papers, replicating various pre-test and 11+ exams. They also include realistic test timings and fully explained answers to help your final test preparation. These papers are designed to improve your accuracy, speed and ability to deal with a wide range of questions under pressure.

# Most unlike

**Have a go** • • • • • • • • • • • • • • • • • • • • • • • • • • •

1 Look closely at this set of pictures. Write **two** features that they have in common. Can you spot a difference that makes one of the pictures stand out as being slightly different from the others? Circle the letter beneath this picture.

   a             b             c             d             e

Feature 1 in common: _____

Feature 2 in common: _____

2 This time write **three** features that this set of pictures have in common. Look at shapes, shading, line styles and angles. Can you spot a difference that makes one of the pictures stand out as being slightly different from the others? Circle the letter beneath this picture.

   a             b             c             d             e

Feature 1 in common: _____

Feature 2 in common: _____

Feature 3 in common: _____

3 In the next set of pictures, look carefully at how shapes fit together and where shading is used. The odd one out is **d**. Can you explain why?

   a             b             c             d             e

**d** is the odd one out because: _____

4 This time the odd one out is **c**. Can you explain why?

   a        b        c        d        e

> Look carefully at the number of lines, shapes and shading.

**c** is the odd one out because: _____

## Test yourself

Look at these sets of pictures. Identify the one that is **most unlike** the others. Circle the letter beneath the correct answer. For example:

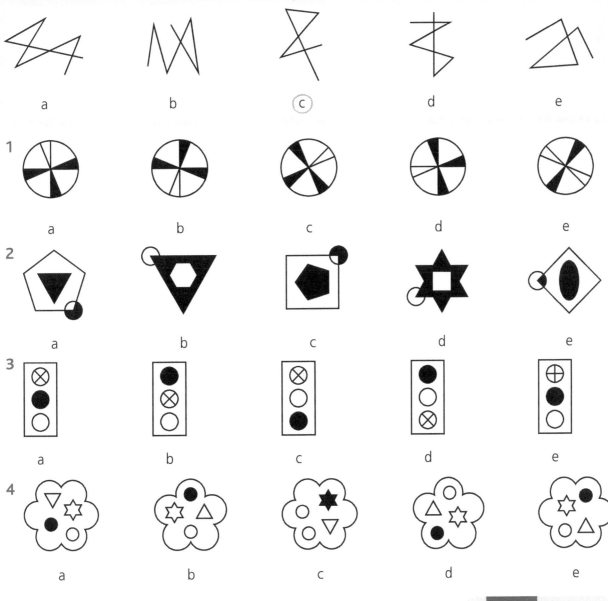

a          b          ⓒ          d          e

1

a          b          c          d          e

2

a          b          c          d          e

3

a          b          c          d          e

4

a          b          c          d          e

Score ☐ / 4

## Try it out

Complete the following pictures so that they all have things in common except option **e**, which should be different from the other four pictures in some way. For example, you could add small shapes, more lines or shading.

    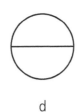    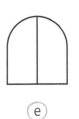

a          b          c          d          ⓔ

9

# Matching features 1

 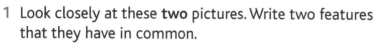

1 Look closely at these **two** pictures. Write two features that they have in common.

Feature 1 in common:

_____

Feature 2 in common:

_____

2 This time write **three** features that these two pictures have in common.

Feature 1 in common:

_____

Feature 2 in common:

_____

Feature 3 in common:

_____

3 This time look for features that the first two pictures have in common and complete the third picture so it matches them using just **two** of the features you spotted.

> To give you a clue, look at the shading and the position of the smaller shapes.

4 The first two pictures below have a number of features in common. But this time there is also a unique feature in each that makes them different from each other. Complete the third picture using **three** of the features that the first two pictures have in common.

> Look at shapes, shading and angles.

5 Write the features that the first two pictures have in common.

Features in common:

_____

_____

## Test yourself

Look at the first two pictures and decide what they have in common. Then select one of the options from the five on the right that belongs in the same set. Circle the letter beneath the correct answer. For example:

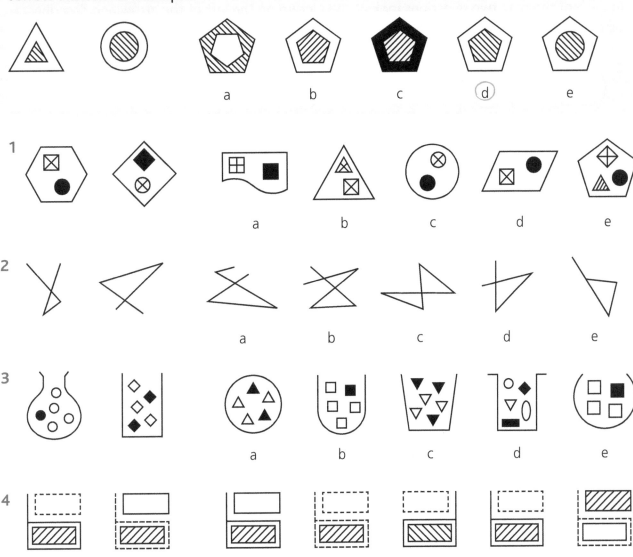

## Try it out

Complete the following question so that answer option **c** has something in common with the first two pictures while the other four options are different in some way. Try not to make your answer too easy.

a      b      c      d      e

# Applying changes 1

In each of the next two questions look at the picture on the left of the arrow and describe what has happened to it in order to make the second picture.

1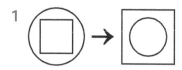

Change that has happened:

_____

2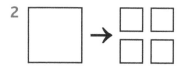

Change that has happened:

_____

In the next two questions, again look at the first picture and decide what changes have happened to create the second picture. Then apply the same changes to the third picture and draw what would happen to it in the box on the right. Keep an eye on shading.

3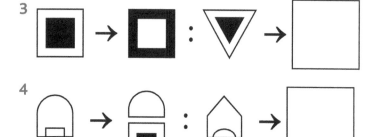

4 

5 Write what changes have taken place between the first and second picture. Then apply those changes to the third picture to work out which of the five options is correct. Circle the letter beneath it.

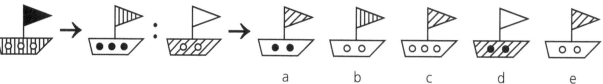

a    b    c    d    e

Changes taking place:

_____

_____

## Test yourself

Look at the two pictures on the left connected by an arrow. Decide how the first picture has been changed to create the second. Now apply the same rule to the third picture and circle the letter beneath the correct answer. For example:

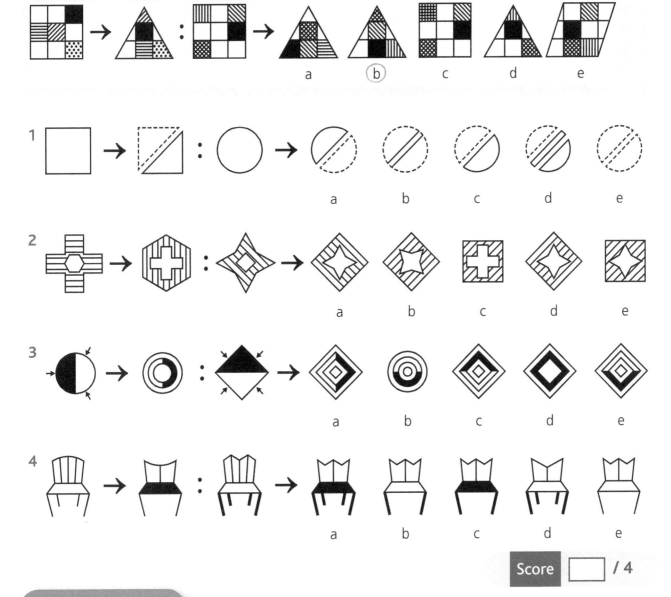

## Try it out

Have a go at creating a question yourself. Complete the first picture and then change some elements and draw the new picture in the first empty box. Next, draw a new picture in the second empty box and ask a friend or parent to answer the question by filling in the third empty box.

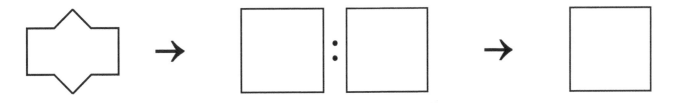

# Matching 2D and 3D shapes 1

1 The two pyramids on the right have been formed from the flat net on the left. The centre triangle of the net is the base of the pyramids. In the second pyramid a face has been left blank. Draw the missing face.

 = 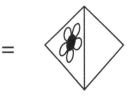 and

2 This time draw the missing faces on both of the pyramids.

 =  and

The next two nets create cubes, like a dice, when folded. Draw the missing pattern that should appear on the top of each cube.

3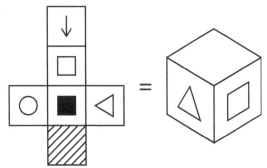

4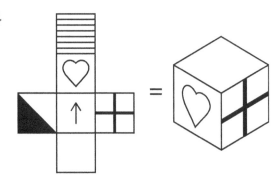

5 Now draw the patterns that should appear on the two visible sides of the cube that are currently blank.

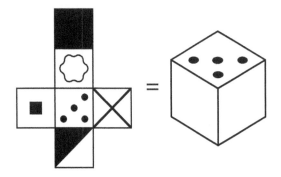

## Test yourself

Find the cube that can be made from the net shown on the left. Circle the letter beneath the correct answer. For example:

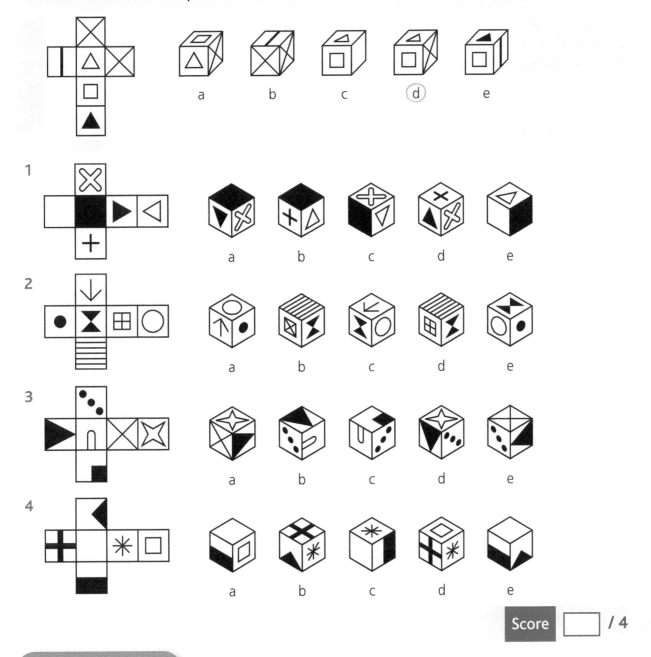

1

2

3

4

Score [ ] / 4

## Try it out

Draw a pattern on each square of the net. Then complete the two cubes by drawing the patterns on the correct faces. Cut up some paper and fold it into place if you need to.

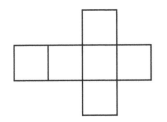 = 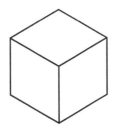 and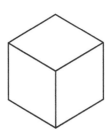

# Matching features 2

1 Look closely at these three pictures. Write two features that they have in common.

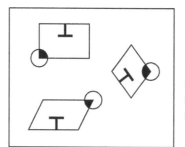

To give you some clues for finding the solution, look at the shading, number of lines and position of the smaller shapes.

Features in common:

(a) _____

(b) _____

2 This time write three features that these three pictures have in common. Look at the shapes, shading and line styles. Can you spot a difference that makes one of the three pictures stand out as being different from the other two?

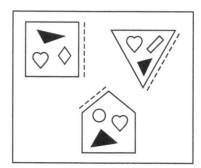

Features in common:

(a) _____

(b) _____

(c) _____

Feature that is different:

_____

3 This time look for features that these three pictures have in common. Complete the fourth picture so it matches them.

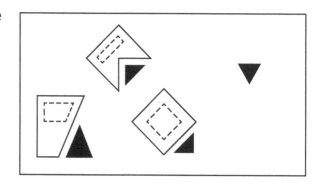

4 These three pictures have a number of features in common, but there are also unique features in each that make them different from each other. Complete the fourth picture using three of the features they have in common.

## Test yourself

Look at the first three pictures and decide what they have in common. Then select the option from the five on the right that belongs to the same set. Circle the letter beneath the correct answer. For example:

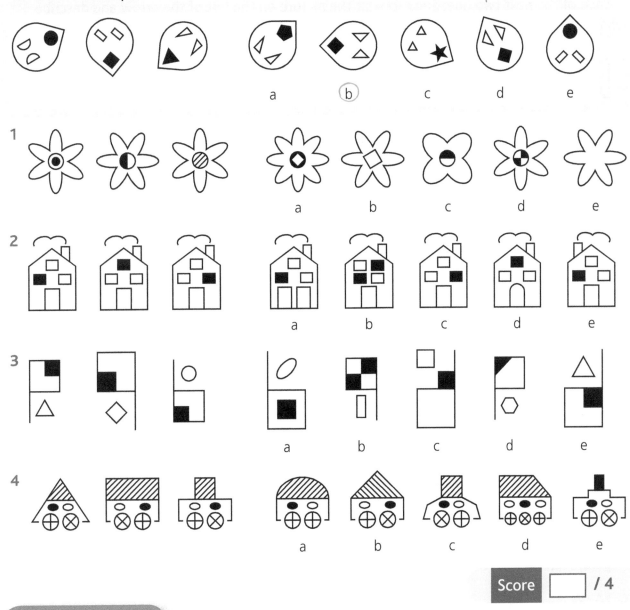

## Score ☐ / 4

## Try it out

Draw patterns in the first three circles so that these first three have some things in common, such as small shapes, similar lines or shading. Then make up some answer options. Design option **b** so it matches the first three circles and is the correct answer. Make options **a, c, d** and **e** different in some way. Try not to make your answer too easy. Ask a friend or parent to try to answer the question.

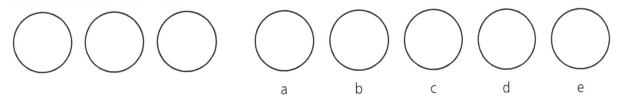

# Applying changes 2

In each of the next two questions, look at the picture on the left of the arrow and describe what has happened to it in order to make the second picture.

1 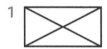 →

Change that has happened:

_____

2  →

Change that has happened:

_____

In the next two questions, look again at the first picture and decide how it has been changed to create the second picture. Then apply the same changes to the third picture and draw what it would look like in the space on the right of the second arrow.

3  :  →

4  →  :  →

5 Work out the changes that have taken place between the first and second pictures and the third and fourth pictures. Explain the changes on the answer line below.

 :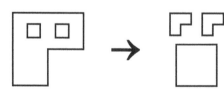

Changes from one picture to another: _____

_____

_____

## Test yourself

Look at the two pictures on the left connected by an arrow. Decide how the first picture has been changed to create the second. Now apply the same rule to the third picture and circle the letter beneath the correct answer. For example:

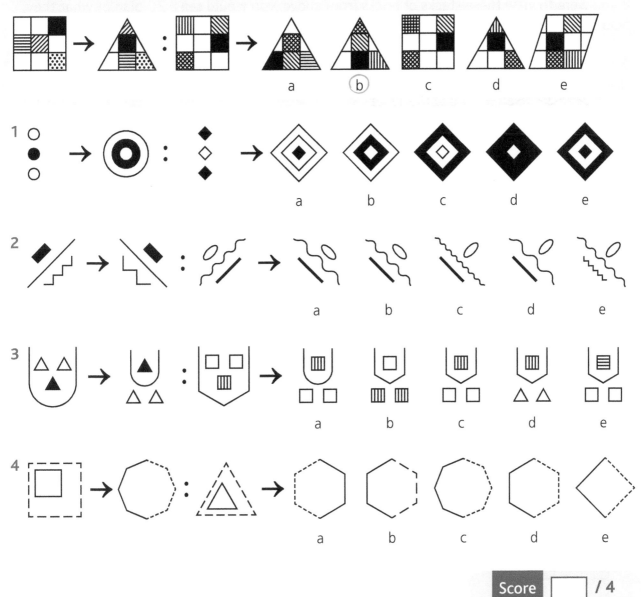

Score ☐ / 4

## Try it out

Complete the first picture by adding small shapes and/or lines with different styles or shading. Then change it in some way to create a second picture. Next, draw a similar picture to the first and change it in the same way as you did with the first picture to create a fourth picture.

# Matching 2D and 3D shapes 2

If you were to view these stacks of bricks from above, you would see a 2D plan of what they would look like. Shade the grid on the right to show what that plan would look like.

1
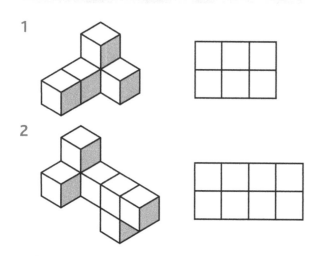

2

This time you are given the 2D plan for a stack of bricks. Circle the letter beneath the stack that matches the plan.

3

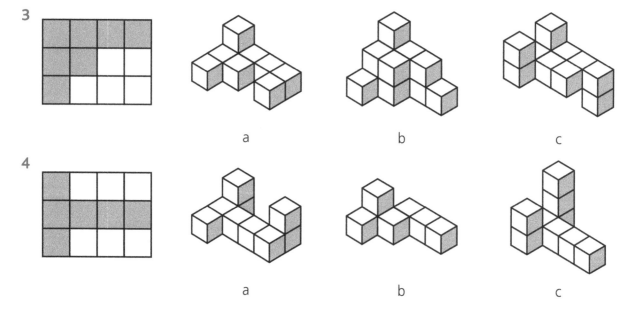

a          b          c

4

a          b          c

5 Now shade in the **two** missing squares on the grid so that it accurately shows the plan of this stack of bricks.

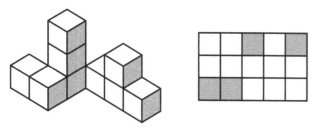

## Test yourself

Which of the answer options is a 2D plan of the 3D picture on the left, when viewed from above? Circle the letter beneath the correct 2D plan. For example:

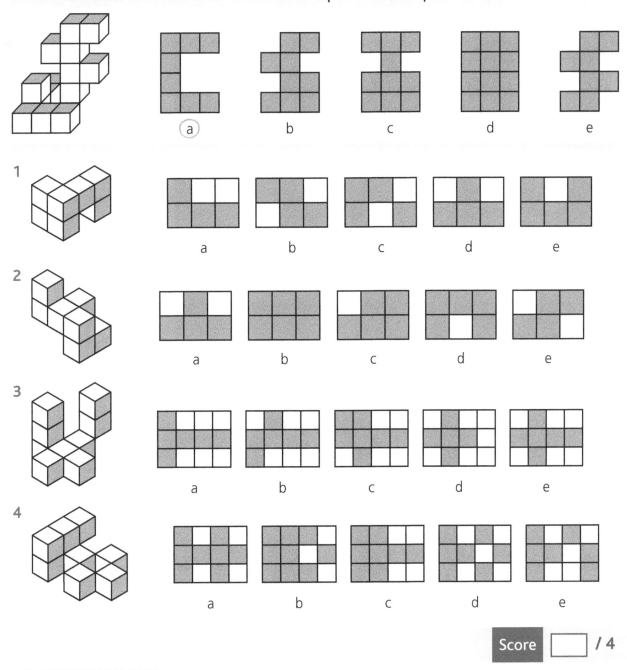

a    b    c    d    e

1

a    b    c    d    e

2

a    b    c    d    e

3

a    b    c    d    e

4

a    b    c    d    e

Score ☐ / 4

## Try it out

Work out how a stack of bricks could be made from this 2D plan and draw it in the space on the right. There are lots of options, so use some bricks if you need help.

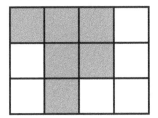

# Following the folds 1

1 Around the edges of the square on the left are some shaped 'flaps'.
When they are folded along the dashed lines back on to the square, they look like the picture on the right. However, one flap is missing. Can you draw it onto the square in the correct place?

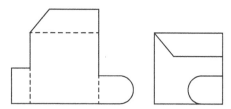

2 This time, the picture on the left shows all the flaps folded onto the square.
Draw where the missing flap will appear on the diagram on the right when it has been folded out.

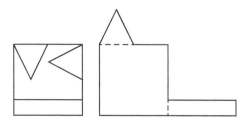

3 Circle the letter beneath the option that shows the diagram on the left when the flaps have been folded **in**.

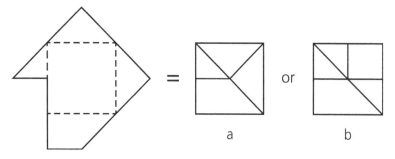

4 Circle the letter beneath the option that shows the diagram on the left when the flaps have been folded **out**.

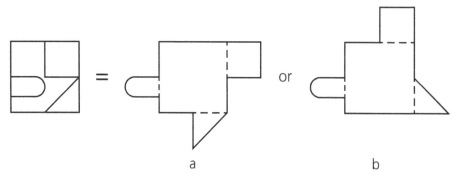

## Test yourself

Identify the diagram which shows how the plan on the left will appear when it is folded in along the dashed lines. Circle the letter beneath the correct answer. For example:

Score [    ] / 4

## Try it out

One square and one triangular flap have been drawn onto a blank square. Draw two possible plans to show how the flaps could be folded out.

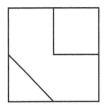 =  or

# Matching a single image 1

## Have a go

1 The picture on the left has been turned upside down (rotated 180°) and placed in the box on the right, but two shapes are missing. Draw the missing shapes in the correct positions in the box.

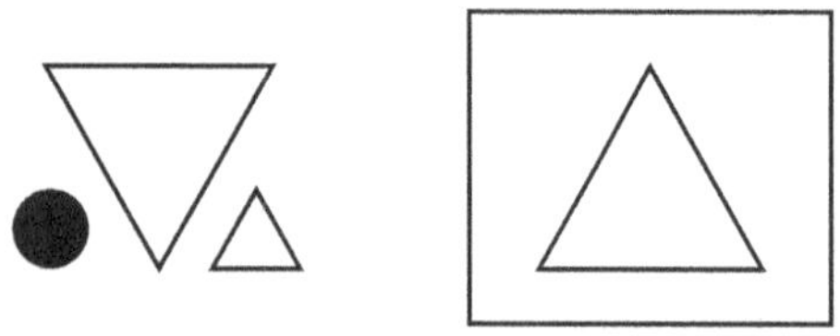

2 The house on the left has been tipped onto its side. Complete the drawing in the box on the right.

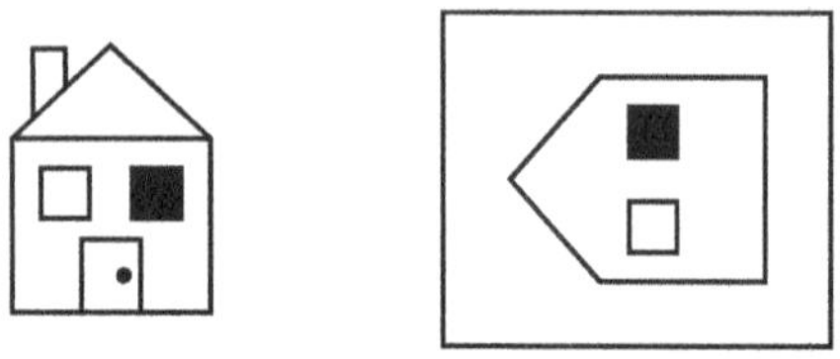

In the next two examples, the picture on the left has been rotated. One of options on the right matches it after it has been rotated. Circle the letter beneath the correct option. If you need to check your answer, turn the page to see which one matches the first picture.

3

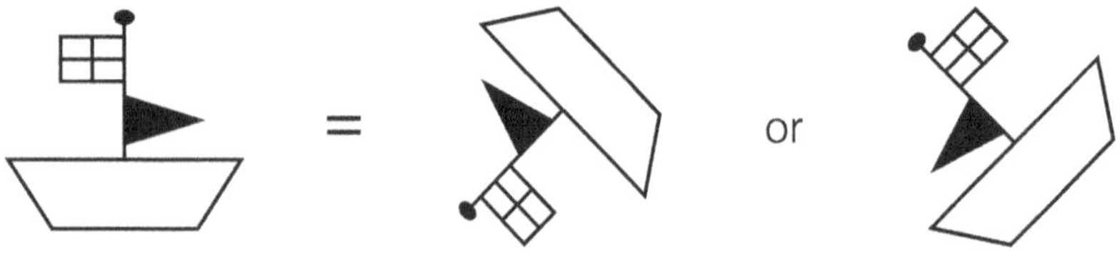


a b

4

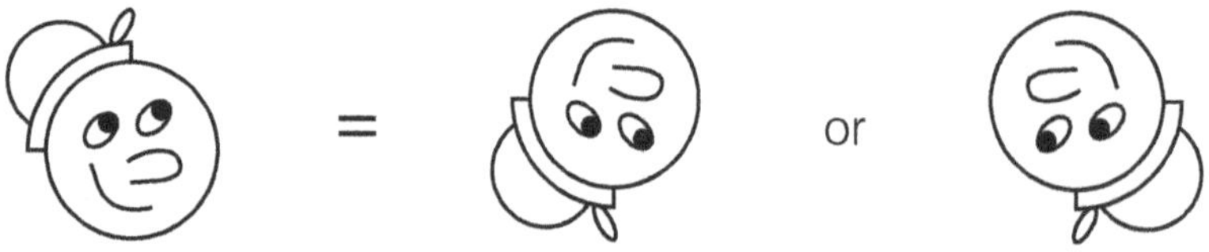


a b

5 Work out which of the two options, **a** or **b**, matches the first picture. Then explain what clues gave it away.

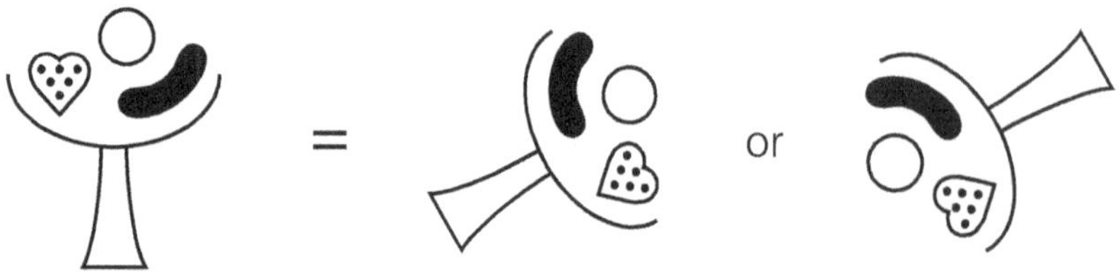


a b

Clues: ______________________________________________

______________________________________________

# Answers

Please note that all questions are worth one mark unless stated otherwise in brackets.

## A Numbers, shapes and relationships

### Most unlike (page 8)

#### Have a go

1   Accept any two of the following features in common: all have the same line style; all have a circle on the end of a line; all have connected lines.
    d   is the odd one out because it has five connected lines, not four.

2   Accept any three of the following features in common: all have ellipses; all have two shapes inside the ellipses; one internal shape is shaded; all ellipses are at the same angle. Single or double external line style is given to distract you from the answer.
    a   is the odd one out because it has a small ellipse instead of a circle inside.

3   d   is the odd one out because the internal shapes in the others are the same as the outline of the top of the outer shape.

4   c   is the odd one out because the other 'trees' each have six branches but c has seven.

#### Test yourself

1   e   number – each circle has three black sectors except e, which has two black sectors
        In common: shape – all circles, same number of sectors inside; shading – some sectors are black, some are white

2   e   shading – (a) each shape consists of a large shape shaded either black or white with a small shape inside with the opposite shading, (b) each shape has a small circle overlapping one corner of the large shape and the shading of the outside part of this small circle matches the shading of the small internal shape except in e
        Distractors: shape – the shapes used are not relevant as long as they are different

3   e   shading/position – each of the pictures contains three circles with solid black shading, white shading and a cross but the cross in e is in a different position
        In common/distractors: shading – each picture contains three circles with the same shading: black, white and a cross; line style – all the figures have the same line weight and style; position – the order of the circles is a distractor

4   c   shading – each cloud contains the same four small shapes and one of the circles is shaded, except for c where the star is shaded instead
        In common/distractors: number – (a) clouds are either five- or six-sided, (b) there are four small shapes inside each cloud; position – the position of the shapes is random

#### Try it out

Addition of lines, shapes or shading to create shapes that are similar, but e must be different from the other pictures.

### Matching features 1 (page 10)

#### Have a go

1   Accept any two sensible answers, which could include the following features in common: both have a large circle with a small black circle inside and a square overlapping the large circle.

2   Accept any three of the following features in common: both have a small tube shape outside the large shape; both large shapes are divided in half; half is shaded with the same stripes; both have a small right-angled triangle in the unshaded half of the large shape.

3   The completed picture should include any two of the following features: a small black square in a corner; a small shape overlapping the opposite corner; a dashed line around the corner of the square to the right of this overlapping shape.

4   The completed picture should include three of the following features: two eye shapes; wavy lines around the outside of the circle; two triangles outside the circle; small triangles inside the circle.

5   Features in common for first two pictures: large outer circle; circle divided into four segments; one segment has striped shading; the lines between the segments are all curved. d is the most like these (note that the stripes in b go in the opposite direction).

#### Test yourself

1   d   shading – one small shape has a cross and one is black; shape – inner shapes are always a square and a circle
        Distractors: shading – the shape that is shaded black or with a cross is unimportant; shape – the outer shape is unimportant

2   d   number – each picture has three lines; shape – two lines cross to make a triangle
        Distractors: length – the length of the lines is unimportant

3   b   shape – (a) each large shape has an open top, (b) the small shapes in each picture are identical; number – each picture contains a large shape and five small shapes
        In common/distractors: shape – (a) the large shape is unimportant as long as it has an open top, (b) each picture contains five identical small shapes; shading – the small shapes can be either white or black, but the number of small shapes shaded is a distractor

4   d   line style/position – both pictures are identical except that the line style of the two outer shapes alternates
        In common: shading – the upper rectangle is unshaded and the lower one is striped

#### Try it out

Addition of lines, shapes or shading to create pictures that are similar but different. Option c must match the first two pictures and so should have three identical internal shapes with one shaded black.

### Applying changes 1 (page 12)

#### Have a go

1   The two shapes have swapped places.
2   The large shape has been split into four identical small shapes.
3   The shapes stay the same but the shading is reversed. The missing picture should show a white triangle inside a black triangle, both pointing downwards.
4   The large shape splits horizontally and the shading of the small shape at the bottom swaps. The missing picture should show a triangle at the top and a square beneath; the oval shape should be white outside the square and striped inside the square.
5   e – shading of flag moves to portholes; shading of boat moves to flag; orientation of boat stays the same

1  c  **shape** – the second shape is the first shape split bottom left to top right; **line style/position** – the upper of the two shapes changes from a solid to a dashed line style

2  d  **position/shape** – (a) the two shapes swap places, (b) the inner shape from the first picture rotates 90° when enlarged; **shading** – shading remains in the larger shape but it rotates 90°

3  e  **number** – the number of arrows dictates how many concentric shapes are in the answer; **shading** – the shading in the first shape moves to the opposite side of the half shape in the second shape inside the concentric set

4  c  **shape** – the shape at the top of the chair turns upside down; **number** – (a) note that the number of bars reduces from three to one, (b) the number of legs reduces by two; **shading** – chair seat changes to black; **line style** – style of chair legs remains the same

## Try it out

Addition of lines, shapes and shading to create pictures that can be transformed.

## Matching 2D and 3D shapes 1 (page 14)

### Have a go

1  The triangle with the striped shading should be added to the blank face. The base of the triangle should be on the bottom edge.

2  The solid black horizontal line should be added to the blank face of the first pyramid and a triangle should be added to the blank face of the second pyramid.

3  A black square should be added.

4  Striped shading should be added. The lines should be parallel to the edge above the heart.

5  A black triangle should be added to the top of the left face and a cross should be added to the right face.

### Test yourself

1  a  Flat edge of black triangle along edge of black face. Open cross between these two faces.

2  d  The lines on the striped face are parallel to the base of the black triangle. The pattern inside the square matches that on the net.

3  d  The extreme ends of net must meet so the flat edge of the triangle is against the star face. If the triangle was flat to the base, then the three dots would have to be on its left. Turn it over and they are on its right.

4  c  When folded you can see that the star will be on the opposite side of the cube from the black cross. This means answer options **b** and **d** cannot be correct. Again, when folded you can see that the side with a triangle will be opposite the half-shaded side so **e** is incorrect. Likewise **a** is wrong because the blank face will be opposite the small square.

### Try it out

Copy the net you have shaded onto squared paper. Cut it out and fold it to make cubes and then complete the two blank cubes.

## Matching features 2 (page 16)

### Have a go

1  Accept any two of the following features in common: four-sided large shape; T-shape inside large shape; circle on a corner; the inner part of the circle where it overlaps the large shape is shaded.

2  Accept any three of the following features in common: three small shapes inside a large shape; always a shaded triangle; always a heart; dashed line down one side of outer shape.
Features that are different: the large shape has three, four or five sides; the top two large shapes contain a four-sided shape in addition to a heart and a triangle, whereas the bottom large shape contains a circle; the dashed line is different lengths; the triangles differ.

3  The completed picture should include a dashed four-sided shape inside a large shape. At least one side of the larger shape should be parallel to a side of the black triangle.

4  The completed picture could include any three of the following: black eyeballs, two buttons, a hat, ears to match the other figures, a mouth of any shape.

### Test yourself

1  d  **number** – six petals; **shape** – must have a small shaded circle
Distractors: **shading** – the shading of the circle is unimportant

2  c  **number** – three windows; **position** – chimney is in the same position; **shading** – one window is shaded; **shape** – the door is a rectangle
Distractors: **shading** – the choice of window that is shaded is unimportant

3  e  **shape** – (a) the flag always contains a black square, (b) there is always one small shape outside the flag; **position** – the black square is always in one corner of the flag
Distractors: **size** – size of the diagram is unimportant; **position** – whether the flag is upright or upside down is unimportant; **shape** –the small shape outside the flag is unimportant

4  c  **number** – two wheels; **shading** – (a) the top part of the picture is shaded with lines running from bottom left to top right, (b) one wheel has a cross and the other a plus
Distractor: the outer shape of the picture is unimportant as are the two ellipses over the wheels

### Try it out

Addition of lines, shapes or shading to create patterns that can be copied or changed to make **b** the correct answer.

## Applying changes 2 (page 18)

### Have a go

1  The width of the shape has reduced by half.

2  One fewer petals. The petals and centre circle have swapped shading.

3  The upper shape has enlarged and now contains the lower shape. Answer should look like this:

4  The centre line rotates and sits at the base, joining the other two vertical lines. The small inner shapes double in number vertically. Answer should look like this:

5  Answer notes should contain: large shape shrinks and its number equals the number of small inner shapes that were inside it; inner shapes become one large shape with small shapes on top.

1   e   **number** – three identical shapes in each figure; **size** – two of the shapes grow in size; **position** – the small shapes become concentric; **shading** – the top small shape shade becomes both the central and the outer shading of the concentric model

2   a   **number and size** – jagged peaks or curves reduce by one and get larger; **position** – the position of the top shape does not change relative to the line beneath it

3   c   **size** – the large shape reduces in size; **position** – the shaded small shape stays inside the reduced shape while the other two go below it; **shading** – the shading of the small shape inside the larger shape remains unaltered (note that the stripes in option **e** have been rotated)

4   d   **number** – the number of sides on the small and large shapes combine to create a new shape with that number of sides; **position** – long-dashed line of the outer shape becomes a line with shorter dashes on the right side of the new shape only

## Try it out

Additions to the first hexagon should be made and then changes in the second pair must match the first pair.

## Matching 2D and 3D shapes 2 (page 20)

### Have a go

1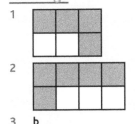

2

3   b

4   c

5   Shaded plan should look like this:

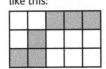

### Test yourself

1   a   The upper row has three cubes in a row going to the right and two cubes going to the left. The second row of cubes sits directly under the top row so the answer is **a** as viewed from above.

2   c   This shape has three rows. One cube on the top row sits above a cube on the second row, which in turn is at the end of a row of three going left. Another cube sits on this row alongside the centre cube and on the right. The third row has two cubes under the first and closest cube of the second row and they go right.

3   e   A stack of three cubes is on the far left. Extending from the right face of the lower cube of this stack is a row of three cubes. There is an additional cube opposite the stack, forming a T-shape when viewed from above. To the right and above the row of three cubes are two cubes touching edges with the upper edge of the far right cube in the row, making the pattern four cubes long.

4   d   On the left is a row of two cubes with three cubes above it. The furthest right cube has nothing below it. Off the right of the two lower cubes is a +-shape of cubes with the central one missing.

## Try it out

Multiple answers acceptable.

## B Position and direction

## Following the folds 1 (page 22)

### Have a go

1   On the bottom left of the square is an additional rectangle as if the outer flap has been reflected vertically.

2   A triangle should be added to top right corner of the square as if the inner flap has been reflected on vertically.

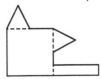

3   a

4   b

### Test yourself

1   b   Each dashed line acts as a mirror for the shape on the outside of the square. So the small square on the left side folds into the bottom-left corner, the triangle on the right side folds onto the square in the bottom-right corner and the two upper shapes fold down onto the top half of the square.

2   d   Each dashed line acts as a mirror for the shape on the outside of the square. So the triangles fold in and meet in the centre of the square, while the small square on the left side folds into the bottom-left corner.

3   e   Each dashed line acts as a mirror for the shape on the outside of the square. This time the entire square is covered with folded-in shapes.

4   a   Each dashed line acts as a mirror for the shape on the outside of the square. This time the entire square is covered with folded-in shapes.

## Try it out

Draw some shapes around a square on squared paper. Cut it out and fold the shapes back onto the square to help you. Some possible answers:

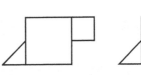

## Matching a single image 1 (page 24)

### Have a go

1 Black circle top **right** of large triangle; inverted small triangle top **left** of large triangle.

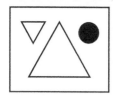

2 The house has been rotated 90° anticlockwise, so elements need to be added accordingly.

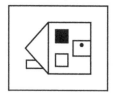

3 a
4 b
5 b Clues should include the position of shapes within the bowl shape.

### Test yourself

1 d **Translation** – 90° clockwise rotation
2 c **Translation** – 180° clockwise or anticlockwise rotation
3 d **Translation** – 90° anticlockwise rotation
4 c **Translation** – 90° clockwise rotation

### Try it out

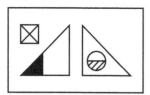

## Translating and combining images 1 (page 26)

### Have a go

1 Item copied across; note proportion of elements and translation of shapes.
2 Item copied across; note proportion of elements and translation of shapes.
3 a Look at the outline of the diagram only.
4 c Look at the outline of the diagram only and ignore shapes that fall inside the overall outline.
5

### Test yourself

1 a The rectangle is narrow. The oval overlaps the top right-hand intersection of the rectangle and centre shape. The centre shape points to the left.
2 d The circle overlaps the straight side of the larger shape. The rectangular indent ends in line with the curve at the top of the larger shape.
3 b The circle is located on the right-hand corner of the larger rectangle. The isosceles triangle points to the left. The smaller rectangle overlaps the larger rectangle and is about half its width.

4 c The right-angle of the triangle is located top left. The arrow points to the left. The square overlaps the side opposite the right-angle of the triangle.

### Try it out

Check the outline is correct by using some tracing paper or greaseproof paper.

## Following the folds 2 (page 28)

### Have a go

1 Four holes: a vertical reflection of the two holes already in place.
Reflection means the holes are reflected across the dashed line created by the fold.

2 Four holes: a 45° diagonal reflection of the two holes already in place.

3 Four holes: a horizontal reflection of the hole that is already in place and then a vertical reflection of the two holes created.

4 Eight holes: a diagonal reflection of the two holes already in place and then a further diagonal reflection of the four holes created.

5 The sheet must be folded twice: once top left to bottom right and a second time through a diagonal, bottom left to top right (it does not matter which diagonal fold is made first). Two holes would need to be drilled to create this pattern.

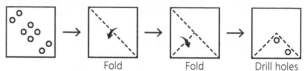

Fold          Fold          Drill holes

### Test yourself

1 e Four holes: horizontal reflection.
2 c Six holes: diagonal reflection. Therefore the orientation of the hearts changes as if they are rotated 90° clockwise or reflected on a mirror angled at 45°.
3 d Twelve holes: horizontal reflection, then vertical reflection.
4 a Eight holes: diagonal reflection, then another diagonal reflection. Therefore the orientation of the hearts changes as if they are rotated 90° anticlockwise or reflected on a mirror angled at 45°.

### Try it out

Check your question by cutting out a square of paper and folding it. Then make holes in the paper safely.

## Matching a single image 2 (page 30)

Matching a single image 2 (page 30)

Have a go

1   Additional curved black flag placed top left

2   Black triangle placed middle left and curved white flag placed bottom right

3   White triangle, top right; black flag with white cross, middle left; black signal-shaped flag, bottom right

4   Black quadrilateral shape with white circle, top left; white triangle, middle right; black flag with white cross outline, bottom left

5   c   Flagpole is a vertical reflection

Test yourself

1   e   When reflected vertically, the black signal shape moves from top left to top right and becomes white. The triangle shape moves from middle right to middle left and becomes black. The black cross on a white background moves bottom left to bottom right and becomes a white cross on a black background.

2   a   When reflected vertically, the curved white flag shape moves from top left to top right and becomes black with white circles. The rectangle shape moves from middle right to middle left and becomes white. The white triangle moves from bottom left to bottom right and becomes black.

3   d   When reflected vertically, the white triangle moves from top left to top right and becomes black. The white triangle with the black cross moves from middle right to middle left and becomes black with a white cross. The black triangle moves bottom left to bottom right and becomes white.

4   b   When reflected vertically, the double white triangle moves from top right to top left and becomes black. The white rectangle with a triangle inside moves from middle left to middle right and becomes black with a black triangle. The white rectangle containing the black triangle moves from bottom right to bottom left and becomes black with a white triangle.

Try it out

Check that your friend or parent has reversed the position and shading of each shape.

## Translating and combining images 2 (page 32)

Translating and combining images 2 (page 32)

Have a go

1   Triangle is at the top right of the grid.

2   Diamond is on the left side halfway down the grid.

3   Target shape is top right of the grid with a horizontal line going through it.

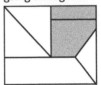

4   Target shape is middle left of the grid with a short line going through it.

5   Target shape is bottom left of the grid, with one line going through it.

Test yourself

1   b          2   d

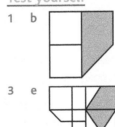

3   e          4   e

Try it out

Check you are correct by cutting the shape from a piece of paper and placing it on top of the grid of lines in the box.

## Working with angles and 2D shapes

1   a
2   d
3   90°, 60°, 45°                                           (3 marks)
4

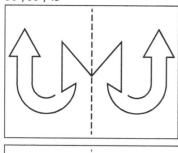

5

                                                            (2 marks)

## Working with 3D shapes

1   23 bricks
2   10 bricks
3

| a | b | c | d | e | f | g | h |
|---|---|---|---|---|---|---|---|
| 22 | 26 | 26 | 26 | 24 | 20 | 20 | 30 |

All the answers are even numbers.                           (9 marks)

## C Codes, sequences and matrices

### Connections with codes 1 (page 36)

#### Have a go

1   **R**    The code stands for the shape; R = triangle
2   **P**    The code stands for the angle shown; P = right-angle
            or 90°
3   **GX**   The codes stand for size and shading
            F and G = size of square; X and Y = shading style
4   **SX**   The codes stand for the number of circles and the
            number of triangles
            R and S = number of circles; X and Y = number of
            triangles

#### Test yourself

1   d    **shape** – the letter R represents the 'V' shape while the
          letter S represents the 'U' shape; **number** – X stands for
          three circles and Y stands for four circles
          Distractors: **position** – the orientation of the shapes and
          circles is unimportant
2   c    **position** – thinking of the black circle as positions on a
          clockface, the letter L represents 9:00, M represents 5:00,
          N represents 1:00; **line style** – G represents a dashed
          outer line around the cloud shape and H a solid outer line
3   a    **shape** – the first letter stands for the shape of the hat,
          with L being curved, M square and N crinkly; **shading** –
          the second letter stands for the shading, with S being
          for vertical stripes and R for diagonal stripes
          Distractors: **shading** – the shading of the headband is
          unimportant

4   d    **number** – the first letter stands for the number of petals,
          with F standing for six petals and G for five; **position** –
          the second letter stands for the position of the dashed
          line, with X being a dashed line around a white petal and
          Y around a black petal
          Distractors: **position** – the position of the black and
          white petals is unimportant

#### Try it out

Codes should match across the question.

### Sequences 1 (page 38)

#### Have a go

1   The circle enlarges and shading swaps sides.

2   The black triangle rotates 90° clockwise around the square.

3   An additional circle is added diagonally down to the right.
    The top circle alternates along the sequence and all circles
    alternate shade.

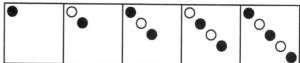

4   An additional petal is added in an anticlockwise direction.
    The petals alternate line style. The centre circle alternates
    shade. The triangle moves down and rotates 90° clockwise.

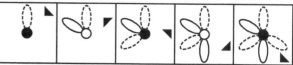

5   There is one less circle in each box. The square in the top-
    right corner enlarges in size and alternates between a × and
    a + inside it.

#### Test yourself

1   c    **size** – the height of the arches alternates; **position** – of
          the triangle alters; **shading** – of the triangle changes; **line
          style** – on tall arches, the dashed line is on the inside
2   a    **number** – of sides on the shape increases by one;
          **rotation** – the circle rotates by 90° around the square;
          **line style** – the additional line in the shape is always
          dashed
3   b    **size** – the size of the square decreases in sequence from
          right to left, so the target box has the smallest square;
          **rotation** – the white circle rotates around the square
          by one corner each time in an anticlockwise direction
          moving in sequence from right to left
          Distractors: the position of the black circle is
          unimportant
4   e    **rotation** – (a) the main diagram rotates 90°
          anticlockwise, (b) the black dot rotates 90° clockwise

#### Try it out

Acceptable sequences include: rotation of the cross shape;
changing its number of arms; shading it; addition of small shapes.

## Matrices 1 (page 40)

### Have a go

1  Horizontally reflected shape from bottom-left corner of matrix. Matrix works on rows.

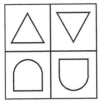

2  90° rotation of shape from top-left corner of matrix; matrix works on columns.

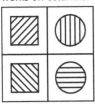

3  Matrix forms a symmetrical pattern, so the bottom right-hand corner is a reflection of the top left-hand corner.

4  Shapes rotate 90° clockwise along their rows.

5  a  The large shape reduces in size and is surrounded by as many circles as sides of the shape. The shading of the circles changes from black to white in the top diagram, so white changes to black in the bottom diagram.

### Test yourself

1  c  **rotation** – the shape in the left column rotates 90° to create the shape in the right column; **size** – the shape loses its left half

2  d  **reflection** – the picture in the right column reflects vertically to create the picture in the left; **shading** – the shading of the two joined shapes swaps over; **position/ size** – the three small shapes change in size and position to create three concentric shapes

3  e  **rotation** – the shapes rotate 90° as they move down the columns; **shape** – the shapes in each column remain the same; **number** – moving down the columns, two additional segments are added

4  b  **size** – the central shape increases in size, moving down the columns; **shape** – the shapes in each column remain the same; **rotation** – the cross rotates 45°, moving down the columns (and is identical on the diagonals)

### Try it out

Accept a row or column format or a symmetrical pattern. Changing shapes, shades and numbers of elements are acceptable, as are rotation and reflection.

## Connections with codes 2 (page 42)

### Have a go

1  **HR**  G and H = large outer shape; R and S = small inner shape; the black bar is a distractor

2  **GLZ**  F and G = direction eyes are looking; L and M = ears or no ears; X, Y and Z = shading of mouth

3  **GRY**  F and G = number of small triangles; R and S = shade style of large triangle; X, Y and Z = orientation of diagram

4  **RMX**  R, S and T = shape; L and M = number of small internal shapes; X, Y and Z = style of outer line

### Test yourself

1  b  **size** – the first letter stands for the size of the shape, with F for small and G for large; **shading** – the second letter stands for the shading, with L for vertical stripes and M for diagonal; **position** – the third letter stands for the position of the black oval, with R for left and S for right

2  d  **number** – the first letter stands for the number of squares, with X for two and Y for three; **line style** – the second letter stands for the style of the water line, with G for wavy and H for jagged; **shading** – the third letter stands for the shading of the squares, with L for black and M for crossed

3  e  **position** – first letter stands for the corner of the shape that is divided, with F for top right, G for bottom right and H for bottom left; **number** – the second letter stands for the number of sections the main shape is divided into, with X for two, Y for three and Z for four; **shading** – the third letter stands for the shading of the triangles, with P for black, Q for white and R for striped

4  d  **shape** (a) – the first letter stands for the outer shape, with W for circle, X for square, Y for triangle and Z for hexagon; **line style** – the second letter stands for the line style of the large shape, with F for dashed, G for thin and H for thick; **shape** (b) – the third letter stands for the small shape, with R for circle, S for triangle and T for oval

Distractors: the small black circles are unimportant. The shading of the small inner shape is unimportant

### Try it out

Codes should match across the question.

## Sequences 2 (page 44)

### Have a go

1  The striped half of the circle rotates 90° clockwise. The white arch rotates 90° anticlockwise.

2  The square rotates 45°, an inner square is added and the previous square becomes shaded.

3  In this alternating pattern, the shaded triangle and circle move round triangles 120° or one segment clockwise.

4    In this alternating pattern, the L-shape rotates 90° clockwise
     while the circle and rectangle rotate 90° anticlockwise.

5    (a)   triangle
     (b)   circle
     (c)   rectangle

Test yourself

1    e    **shape** – the shapes swap between hexagons and
          triangles; **line style** – the line style works in pairs, with
          the first pair of shapes sharing a line style, as do the
          second and third pairs

2    b    **rotation** – (a) the solid black arrow rotates 30°
          clockwise about the centre of the circle (or five minutes
          if the circle is seen as a clock), (b) the open arrow
          rotates 90° clockwise (or 15 minutes); **number** – the
          outer circle increases by one dash each time

3    a    **number** – the jug loses one circle each time but there is
          one square in all jugs; **proportion** – the quantity of the
          fill reduces each time; **shading** – the hatched shading
          reflects each time

4    c    **rotation** – the cube rotates 120° anticlockwise (that is,
          the black circle, large cross and white square all move
          anticlockwise from diamond to diamond)

Try it out

There should be a recognisable sequence and only one possible
answer.

## Matrices 2 (page 46)

Have a go

1    This is an alternate pattern where every other segment is
     identical.

2    90° rotation of pattern.

3    The upper shaded shapes lose their shading and move to the
     diagonally opposite position.

4    There is one of every shape in each row. This is a special
     pattern known as a Latin square where one diagonal
     contains all the same shape while the other shapes are
     distributed in a triangular fashion.

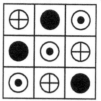

5    b    The black triangle rotates 60° so will be in the bottom-
          right corner. The small shape alternates between a circle
          and a triangle.

Test yourself

1    c    **shape** – the large shape increases in height and narrows
          anticlockwise starting from the triangular cell on the
          left below the centre line; **number** – each cell contains
          one, two or three small diamonds and the number of
          diamonds equals that in its opposite cell; **position** –
          only one diamond of the opposite pairs of cells must be
          inside the large shape

2    e    **number** – (a) one additional circle is added to each
          hexagon, moving clockwise around matrix, (b) one fewer
          lines moving clockwise around matrix; **shading** – the
          new circle is always the black one

3    a    **position** – the first and third shapes in the rows join
          in the centre, but swap position; **shading** – the figures
          swap shading

4    a    **shape** – the shape pattern works in diagonals moving
          top right to bottom left, with the top 'quarter' shape
          being added once to the second row to form a 'half' and
          then again to the bottom row to form 'three-quarters';
          **rotation** – the 'quarter' is always added in a clockwise
          direction so that the shape stays in the same position in
          the cell
          Distractors: the thick lines are irrelevant

Try it out

There are many possible answers although only one of the five
answer options must work in the blank space on the hexagon grid.

## Maths workout 2 (page 48)

### Working with fractions and area

1    **a** and **d**
2    **a** and **b**
3    **a**, **b** and **c**
4    **a**, **b**, **d** and **e**

## Test yourself

The picture on the left is rotated as shown by the arrow. Which answer option shows the picture after the rotation? Circle the letter beneath the correct answer. For example:

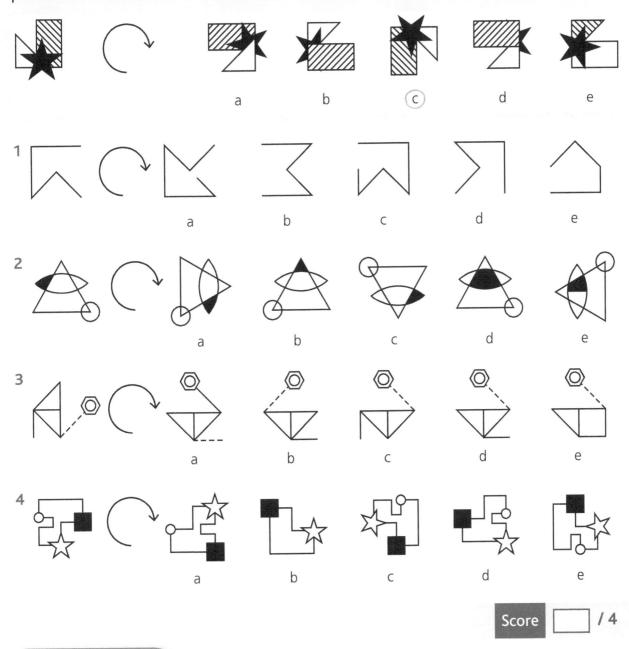

## Try it out

Imagine the picture on the left has been turned upside down, or turned through 180° (**not** reflected as if in a mirror). Redraw it in its new position inside the box on the right.

# Translating and combining images 1

Redraw the picture in the box beneath. Use the grid lines to help you. Be as accurate as you can.

1

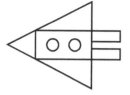

2

In the next two questions, a picture is given on the left. It will only fit perfectly through **one** of the three windows on the right. Circle the letter beneath the correct window.

3

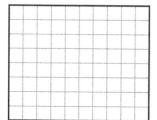

a          b          c

4

a          b          c

5 This time, in the box on the right, draw the hole through which the picture on the left has to pass through. The shading has been added to distract you. You only need to draw the outline of the complete picture.

## Test yourself

Look at the picture on the left. One of the holes on the right matches this picture exactly. Circle the letter beneath the hole that is an exact match. For example:

a   b   ⓒ   d   e

**1**

a   b   c   d   e

**2**

a   b   c   d   e

a   b   c   d   e

**3**

a   b   c   d   e

Score ☐ / 4

## Try it out

Create a question yourself. Draw three overlapping shapes in the space on the left. Then draw **just the outline** of the picture in the box to create a perfect hole for it to fit through.

# Following the folds 2

Imagine a square sheet of paper is folded in half along the dashed line and in the direction of the arrow. This is shown by diagrams 1 and 2 below. Two holes are then punched through the folded paper. When the paper is unfolded, where do you think the holes will appear? Draw them in the blank box.

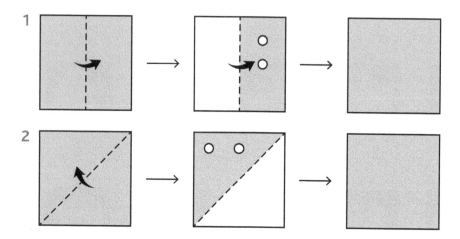

In the next two questions, the sheet of paper has been folded twice, each time along the dashed line and in the direction of the arrow. Draw the holes created when the paper is unfolded.

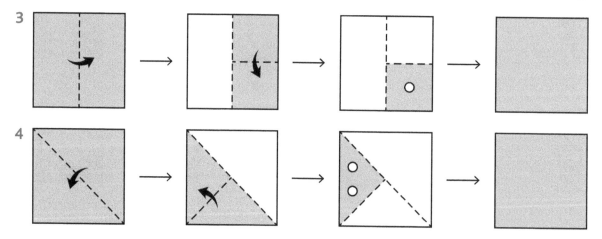

5 This time a drilled sheet of paper is provided on the left. Using the method shown above, draw where folds have taken place and holes have been made to create this drilled sheet of paper.

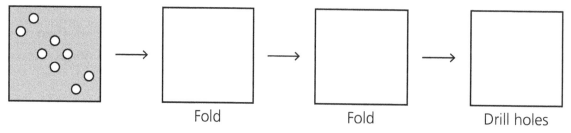

Fold          Fold          Drill holes

## Test yourself

The square given at the beginning is folded in the way indicated by the arrows, and then holes are punched where shown on the final diagram. Identify the answer option which shows what the square would look when it is unfolded. Circle the letter beneath the correct answer. For example:

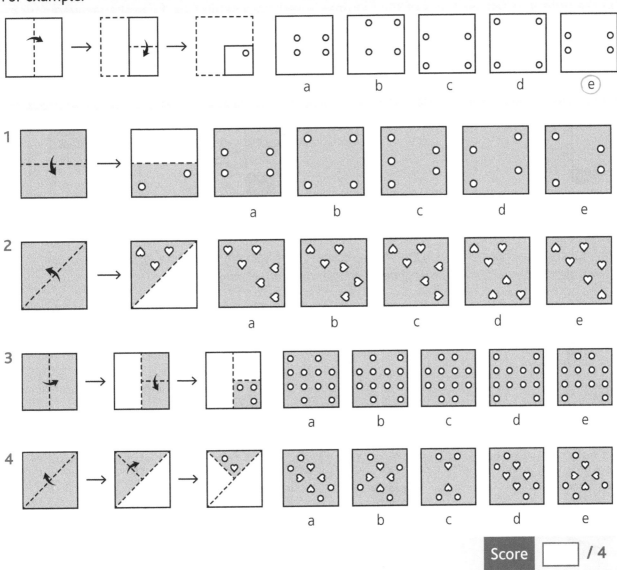

## Try it out

Try creating a question yourself. Draw a dashed line in the first square to represent a fold. In the second square show the folded sheet and add two holes. In the final square, show where all the holes will appear when the sheet is unfolded.

# Matching a single image 2

On the left you can see some shapes connected to a vertical line. These look a bit like flags. On the right you can see parts of the picture viewed from behind, as if the shapes were stuck onto a window and you were looking at them from the other side. Notice that the shading reverses on the other side. Complete the pictures in the boxes to show them viewed from the other side of the window.

1 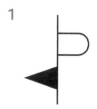   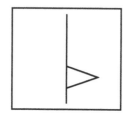   2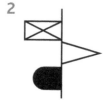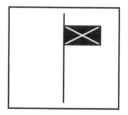

Draw all the missing shapes on the vertical lines, as if you were looking at them from the other side of the window. Remember that their shading changes too.

3    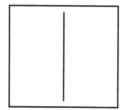   4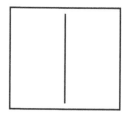

5 Circle the letter beneath the correct view of the picture from the other side of the window.

   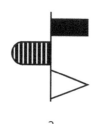      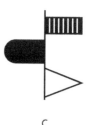

a                    b                    c

## Test yourself

The shapes on the first vertical line are made out of paper that is black on one side and white on the other. It is then pasted onto a window. One of the answer options shows the same pattern viewed from the other side of the window. Remember that the shading reverses when viewed from the other side. Circle the letter beneath the correct answer option. For example:

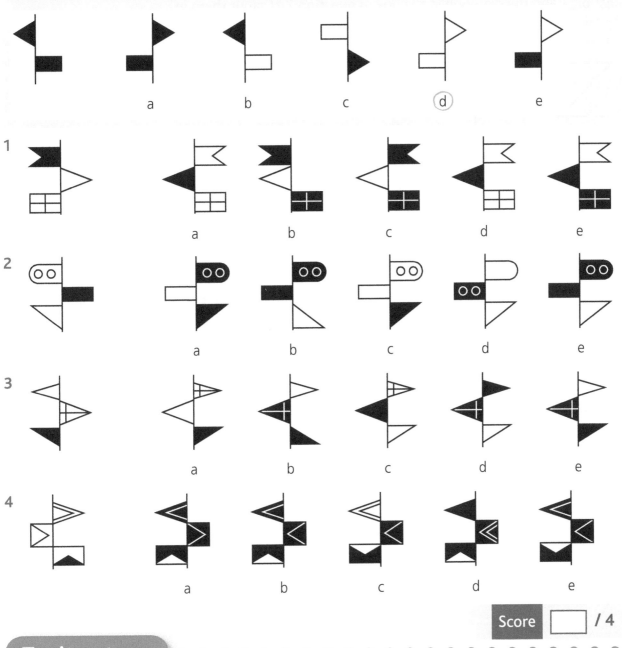

Score [ ] / 4

## Try it out

Draw shapes on the vertical lines below. Put two shapes on the first line and three shapes on the second line. Now ask a friend or parent to draw the pictures in the boxes as if viewed from the other side of a window, with the shading reversed.

# Translating and combining images 2

These questions are all about finding hidden shapes. Can you find the triangle in the box in question 1 and the diamond in the box in question 2? Colour them in.

1

 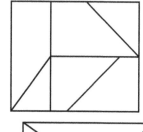

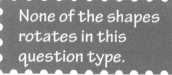
None of the shapes rotates in this question type.

2

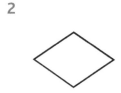 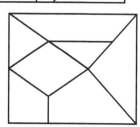

The next three questions are a little trickier. You have to find the shape on the left hidden in the picture on the right. There are lines overlapping the shape you are looking for, but there is only one possible place the shape can be. Colour in the area of the picture that you think matches the shape.

3

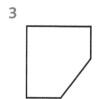 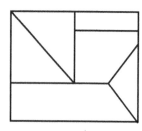

4

 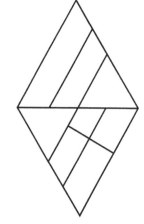

5

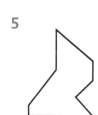 

## Test yourself

The small shape on the left can be found in one of the pictures on the right. It might be made up of one or more pieces. Circle the letter beneath the correct answer option. For example:

Score [  ] / 4

## Try it out

Create a grid of lines in the box to hide the shape on the left. Use rotation and overlapping lines to disguise it.

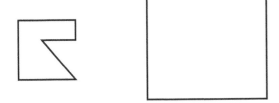

# Maths workout 1

Non-Verbal Reasoning papers include mathematical questions because there are many strands of mathematics that have a direct link with reasoning. The need to identify shapes and patterns, symmetry and rotation are all elements of both maths and Non-Verbal Reasoning. These quick questions will help to sharpen those mathematical skills which will, in turn, help you with Non-Verbal Reasoning questions.

# Working with angles and 2D shapes

1  In which of the following options do the two lines join at 90°?

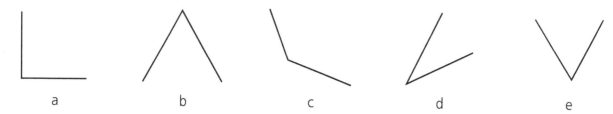

    a             b            c            d            e

2  Which shape has angles that add up to 180°?

    a             b            c            d            e

3  Write the angles on the answer lines below. Choose from: 20°, 45°, 90°, 120° and 60°.

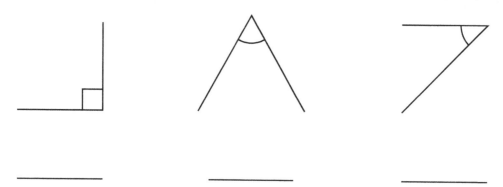

_____        _____        _____

4  Draw an exact reflection of the picture on the left to the right of the dashed line.

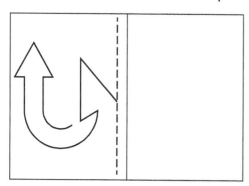

5 Draw an exact reflection of the picture on the left to the right of the dashed line.
  Then reflect **both** pictures beneath the horizontal dashed line.

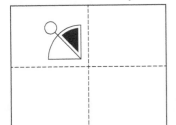

# Working with 3D shapes

How many bricks are in each of the two stacks below?

1

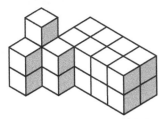

  Number of bricks: _____

2

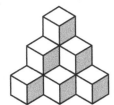

  Number of bricks: _____

3 Here are several stacks of bricks. Each brick has six faces. In many cases the faces touch. Count
  how many faces can be seen from the **outside** of the stack. Don't forget to include the outside
  faces that are hidden from view in these pictures. Write your answers in the table below.

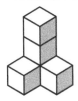

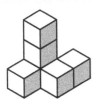

  a              b              c              d

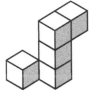

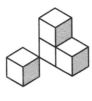

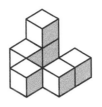

  e              f              g              h

| a | b | c | d | e | f | g | h |
|---|---|---|---|---|---|---|---|
|   |   |   |   |   |   |   |   |

  What do you notice about all the answers? All the answers are: _____

# C Codes, sequences and matrices

## Connections with codes 1

• • • • • • • • • • • • • • • • • • • • • • • •

Look at the shapes with letters beneath. Each letter is a code representing a feature of the shape. Write the code for the fourth shape on the rule beneath it. Write the reason for choosing the code on the answer lines below each question.

**1**

    R           S           T         \_\_\_\_\_

Reason for the code: _____

**2**

  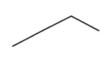 

    P           Q           R         \_\_\_\_\_

Reason for the code: _____

The next two questions contain two-letter codes to represent different features. Write the fourth code on the rule beneath the last diagram. Write your reason for choosing the code on the answer line below each question.

**3**

   FX        FY        GY      \_\_\_\_\_

Reason for the code: _____

_____

**4**

   RX        RY        SY      \_\_\_\_\_

Reason for the code: _____

_____

Each letter represents an individual feature in the picture next to it. Work out which feature is represented by each letter. Apply the code to the picture in the box and circle the letter beneath the correct answer code. For example:

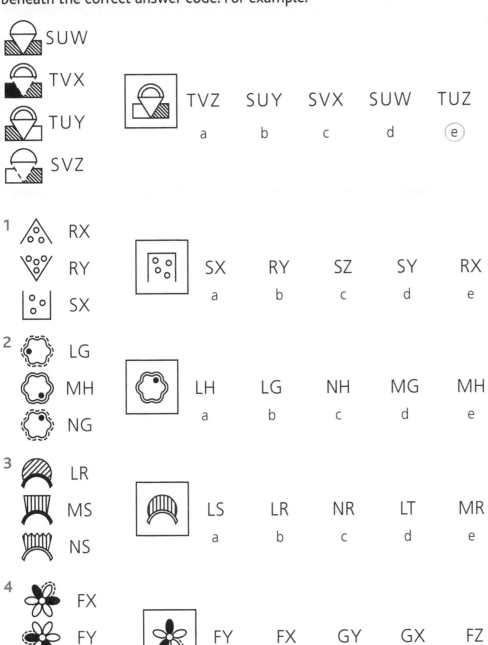

SUW

TVX

TUY

SVZ

TVZ    SUY    SVX    SUW    TUZ
a        b        c        d       (e)

**1**   RX   RY   SX

SX    RY    SZ    SY    RX
a       b       c       d       e

**2**   LG   MH   NG

LH    LG    NH    MG    MH
a       b       c       d       e

**3**   LR   MS   NS

LS    LR    NR    LT    MR
a       b       c       d       e

**4**   FX   FY   GY

FY    FX    GY    GX    FZ
a       b       c       d       e

Score ☐ / 4

Draw five pictures, matching elements of each to two code letters. The first code letter should relate to a shape and the second code letter to a shade. Write the codes beneath the first four pictures. Ask a friend or parent to work out the code for the fifth picture.

# Sequences 1

In these sequences each picture changes in some way from the picture before it. To answer questions 1 and 2, draw the fourth picture in the sequence in the empty box.

1

2

This time the missing picture comes in the middle of the sequence. Draw the missing picture in the empty box.

3

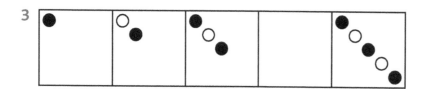

4

5 Describe what is happening in this sequence on the answer line below. What is happening to the square and the circles?

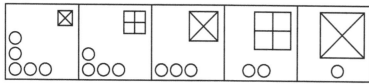

Changes taking place: _____

_____

_____

## Test yourself

The five boxes on the left show a pattern that is arranged in a sequence. Choose the answer option that completes the sequence when inserted in the blank box. Circle the letter beneath the correct answer. For example:

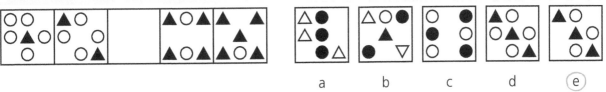

**1**

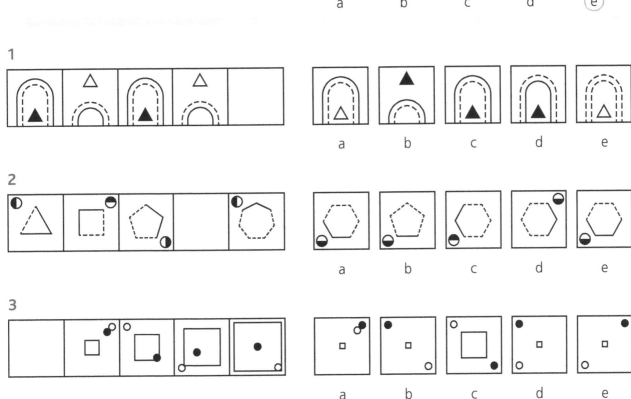

**2**

**3**

**4**

Score ☐ / 4

## Try it out

Make up a sequence of four pictures, starting with the picture shown in the grid below. Ask a friend or parent to draw the fifth picture.

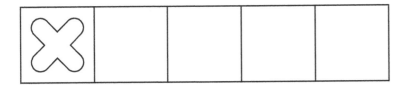

# Matrices 1

Draw the missing patterns in the blank boxes.

1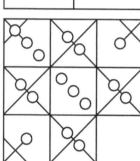

2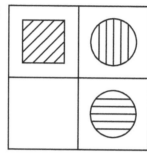

> Look at columns, rows and diagonals to help you complete these patterns.

3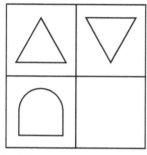

4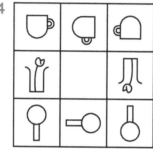

5 Which of the three patterns on the right of the grid fits the blank box? Circle the letter beneath the pattern and explain your choice on the answer line below.

a      b      c

Reason for choice: _____

One of the options on the right completes the pattern in the grid on the left. Circle the letter beneath the correct answer. For example:

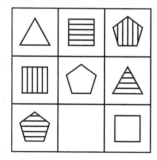                       

(a)     b     c     d     e

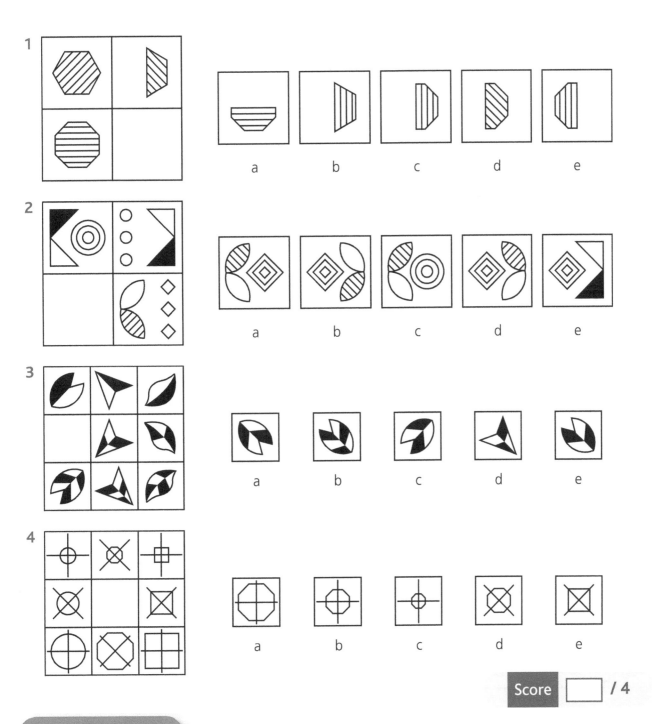

Score ☐ / 4

## Try it out

Try creating a question yourself. Draw a pattern in the grid below where shape, number and shading change in some way. Remember, you can use symmetry, columns or rows to create your pattern. Leave a blank square in the grid and draw five possible answer options in the boxes on the right. Then ask a friend or parent to try to answer the question.

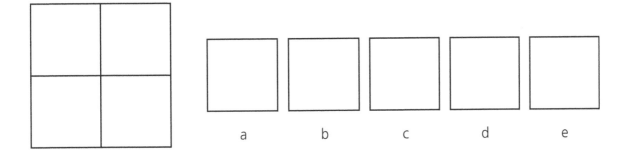

# Connections with codes 2

**Have a go** • • • • • • • • • • • • • • • • • • • • • • • • • • • •

Work out the code for the fourth picture and write it on the rule beneath it. Write the reason for your choice on the answer lines provided.

**1**

GR HS GS _____

Reason for code: _____

**2**

FLX GLY FMZ _____

Reason for code: _____

Work out the code for the fourth picture. Then explain what the code letters stand for on the answer lines provided.

**3**

FRX FSY GSZ _____

F and G stand for: _____

R and S stand for: _____

X, Y and Z stand for: _____

**4** Work out the code for the picture in the box and then explain what the code letters stand for on the line provided.

 RLX

 SMX

 TLY

 SMZ

R, S and T stand for: _____

L and M stand for: _____

X, Y and Z stand for: _____

_____

## Test yourself

Each letter represents an individual feature in the picture next to it. Work out which feature is represented by each letter. Apply the code to the picture in the box and circle the letter beneath the correct answer code. For example:

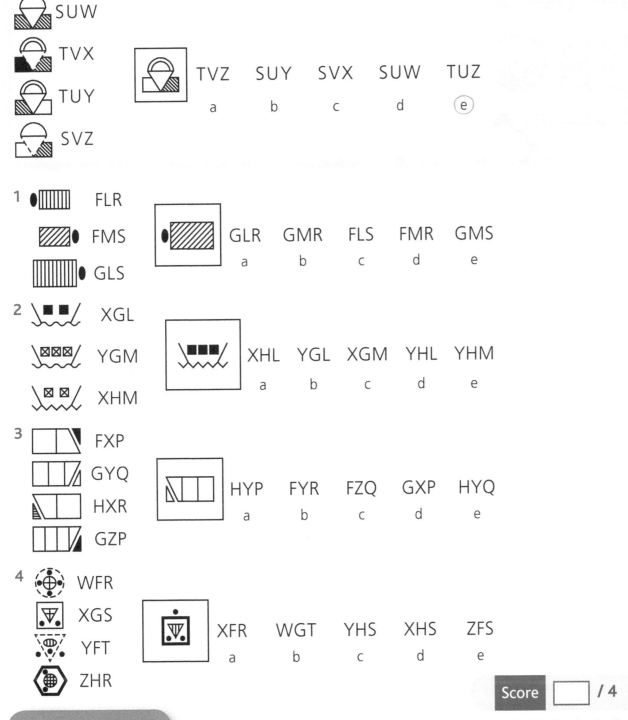

| | SUW |
| | TVX |
| | TUY |
| | SVZ |

| | TVZ | SUY | SVX | SUW | TUZ |
| | a | b | c | d | (e) |

1 | | FLR
| | FMS
| | GLS

| GLR | GMR | FLS | FMR | GMS |
| a | b | c | d | e |

2 | | XGL
| | YGM
| | XHM

| XHL | YGL | XGM | YHL | YHM |
| a | b | c | d | e |

3 | | FXP
| | GYQ
| | HXR
| | GZP

| HYP | FYR | FZQ | GXP | HYQ |
| a | b | c | d | e |

4 | | WFR
| | XGS
| | YFT
| | ZHR

| XFR | WGT | YHS | XHS | ZFS |
| a | b | c | d | e |

Score [    ] / 4

## Try it out

Now try to create a question yourself. Draw five pictures. Match features of each picture to two code letters and write the codes under the first four pictures only. The first code letter should relate to a shape, the second code letter to shading and the third code letter to a variation in size. Ask a friend or parent to work out the code for the fifth picture.

# Sequences 2

In these sequences each picture changes in some way from the picture before it. Complete the final picture in each sequence.

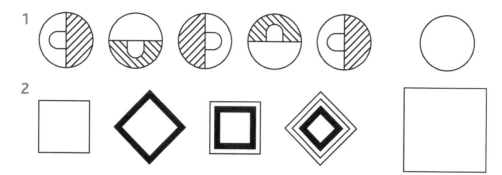

For questions 3 and 4 draw the missing picture in the sequence.

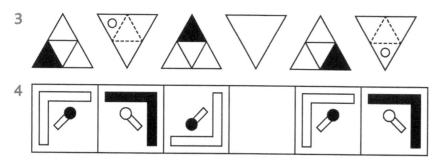

5 Look at all the elements in this sequence. Then answer the questions about it on the answer lines below.

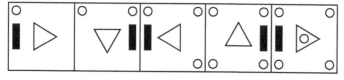

(a) Which shape rotates 90° clockwise? _____

(b) Which shape increases by one each time? _____

(c) Which shape alternates its position? _____

## Test yourself

The five boxes on the left show a pattern that is arranged in a sequence. Choose the answer option that completes the sequence when inserted in the blank box. Circle the letter beneath the correct answer. For example:

        a        b        c        d        (e)

The following questions have either a space or an outline for where the missing picture in the sequence belongs.

**1**

a        b        c        d        e

**2**

a        b        c        d        e

**3**

a        b        c        d        e

**4**

a        b        c        d        e

Score ☐ / 4

**Try it out** • • • • • • • • • • • • • • • • • • • • • • • • • • • • •

Draw five circles. Draw a sequence of four patterns in them and leave one blank. In five other circles, suggest some answer options and see if a friend or parent can guess which is the missing pattern.

# Matrices 2

Draw the missing patterns in the blank spaces.

1

2

3

4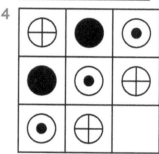

5 Which of the three options on the right fits the blank space in the grid? Circle the correct answer option.

a          b          c

**Look for clues in shapes, reflection and rotation.**

**Test yourself**

One of the options on the right completes the pattern in the grid on the left. Circle the letter beneath the correct answer. For example:

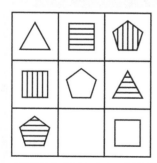              

(a)          b          c          d          e

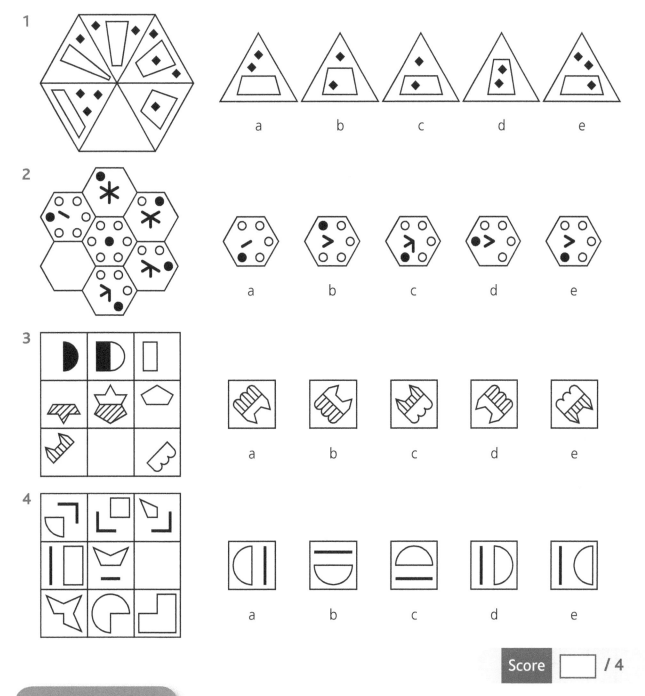

1

2

3

4

Score ☐ / 4

## Try it out ● ● ● ● ● ● ● ● ● ● ● ● ● ● ● ● ● ● ● ● ● ● ● ● ● ● ●

Now try creating a question yourself. Draw a pattern in the grid below. Remember, you can use symmetry, shapes, reflection and rotation to create the pattern. Leave one hexagon in the grid blank and give five possible answer options in the hexagons on the right. Ask a friend or parent to try to answer your question.

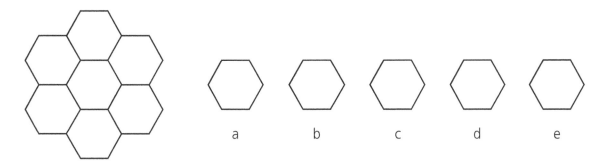

a          b          c          d          e

# Maths workout 2

Many Non-Verbal Reasoning problems make use of mathematical skills and knowledge, so these pages contain some questions and puzzles to strengthen your mathematical skills, vocabulary and ideas. Keeping your maths skills sharp will help you to solve Non-Verbal Reasoning questions more quickly!

## Working with fractions and area

1 Which shapes below have $\frac{1}{4}$ shaded? There may be more than one!

a              b              c              d              e

2 Which shapes below have $\frac{1}{3}$ shaded? There may be more than one!

         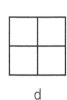   

a              b              c              d              e

3 Which **three** sets of squares below can be fitted exactly into the grid? Circle the letters beneath the **three** correct sets of squares.

a              b              c              d

4 Which **four** sets of squares below can be fitted exactly into the grid? Circle the letters beneath the **four** correct sets of squares.

a              b              c              d              e

Score ☐ / 4